Genetic Engineering

Ron Fridell

LERNER PUBLICATIONS COMPANY
MINNEAPOLIS

Lerner Publications Company
A division of Lerner Publishing Group
241 First Avenue North
Minneapolis, Minnesota 55401 U.S.A.

Website address: www.lernerbooks.com

Library of Congress Cataloging-in-Publication Data

Fridell, Ron.
    Genetic engineering / by Ron Fridell.
        p.   cm. — (Cool science)
    Includes bibliographical references and index.
    ISBN-13: 978–0–8225–2633–9 (lib. bdg. : alk. paper)
    ISBN-10: 0–8225–2633–6 (lib. bdg. : alk. paper)
    1. Genetic engineering—Juvenile literature. 2. Biotechnology—Juvenile literature. I. Title. II. Series.
    QH442.F745 2006
    660.6'5—dc22   2004022764

Manufactured in the United States of America
1 2 3 4 5 6 – BP – 11 10 09 08 07 06

Table of Contents

Introduction

Imagine inventing life-forms never before seen on planet Earth. Think of corn plants that fight back when insects attack or of goats that give spider silk in their milk. What about bacteria that make medicines for sick people? Pigs that grow human body parts might sound like a good idea. What can you imagine?

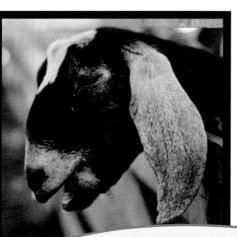

Genetic engineers have created goats that can make the same silk in their milk that this orb weaver spider makes.

This scientist uses a special microscope that allows him to make changes to cells.

If you are a genetic engineer, you are a scientist who is working to bring ideas like these to life. In fact, all of the examples just mentioned are real. In laboratories around the world, genetic engineers think of changes to make to plants or animals. They perform experiments to make them happen.

This work takes huge amounts of money, energy, and patience, but it produces amazing results. Right this minute, scientists are working on hundreds of promising new projects. Genetic engineering can help feed the hungry and heal the sick, and may even make the planet a cleaner and safer place to live.

chapter 1

How Genetics Works

The key to genetic engineering is a component (part) of living things called a genome. One of its discoverers, biologist Francis Crick, called it "the secret of life."

A genome holds a full set of nature's instructions for growing a specific plant or animal and keeping it alive. By changing these instructions, scientists can change the way living things grow and behave. And that's what genetic engineering is all about—changing living things by changing their genomes.

James Watson (left) and Francis Crick (right) were both scientists who studied genetics. They worked together to figure out the structure of chemicals in the genome.

Before we can change a genome, we need to find it. So let's get moving. We'll need to shrink ourselves until we're smaller than the period at the end of this sentence. Then we'll climb inside an imaginary miniature submarine and plunge inside a human being.

Once inside, we don't have far to look. Nearly every cell of every person has an identical copy of that person's genome. So let's go inside the nearest cell to its liquid center, the nucleus. Inside the nucleus are twenty-three pairs of rod-shaped structures. The rod-shaped structures are chromosomes. Chromosomes are made of tightly coiled threads called deoxyribonucleic acid (DNA). Separate chemical units are arranged along the DNA threads. Each separate unit of chemicals is a gene.

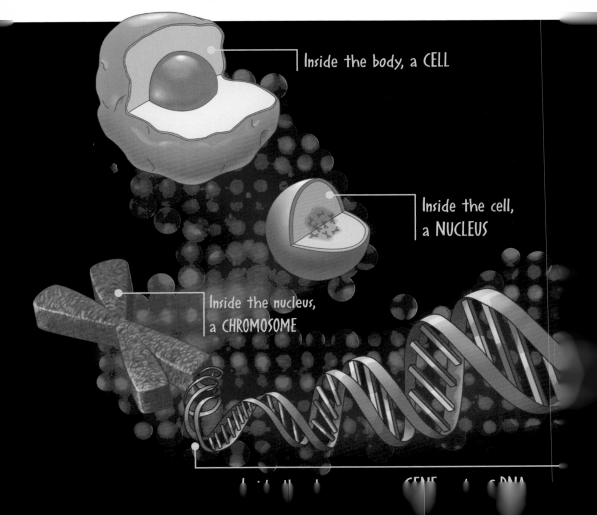

Inside the body, a CELL

Inside the cell, a NUCLEUS

Inside the nucleus, a CHROMOSOME

GENE   DNA

This is it. We have arrived. We are looking at the human genome, the secret of life. The strings of DNA have all of the information needed for your cells to live and grow. And the DNA in one of your cells is the same as the DNA in all of your other cells.

A genome is all of the genes in all of the chromosomes of an organism. Every gene tells a living organism how to make or do something different. Genes help determine the color of a person's eyes, the shape of a butterfly's wings, the length of a jackrabbit's ears, how sour a lemon tastes, how sweet a rose smells, how fast a person's heart beats, and everything else about the physical makeup of living things. Genes also affect the length of time a tomato takes to ripen, the amount of milk a cow produces, and how likely a person is to get certain diseases. Genes work by instructing cells to make proteins. Proteins are building blocks that make up cells.

Twenty-three Pairs from Where?

Humans get one set of twenty-three chromosomes from one parent and one set of twenty-three chromosomes from the other. Together, these add up to forty-six chromosomes, or twenty-three pairs. Parents' chromosomes are mixed together in a different way every time they have a child. That's why each person's genome is different from everyone else's, except for identical twins (right).

Same Gene, Different Color

Let's say you have brown eyes and a friend has blue eyes. Does that mean you have different genes? Yes and no. You both have the genes that carry instructions for eye color. Those genes are located in the same places on the same chromosomes in both of your genomes. But your genes call for brown eyes, and your friend's genes call for blue.

To find the genome of any plant or animal, you would look inside the nucleus of that organism's cells. But you would see something different from your own human genome. Other genomes are not exactly the same as human genomes—other organisms have different numbers of chromosomes and different numbers of genes.

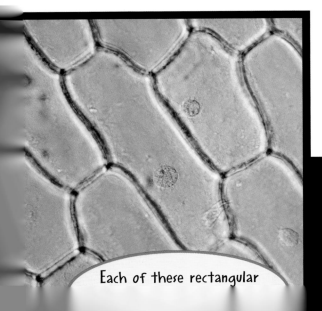

Each of these rectangular

FUN FACT!

In the 1660s, English scientist Robert Hooke named cells after the small rooms where monks lived

Since the early 1980s, scientists have engineered (changed) the genomes of plants, animals, and even a few humans. Every day we eat food that has been engineered. You may even have an engineered pet one day. Let's look at a few of the amazing things genetic engineers have done.

This goldfish has ninety-four chromosomes!

Strings of DNA

A single strand of DNA is too small to see with the naked eye. But you can see big clumps of DNA without a microscope. In 1868 Swiss scientist Friedrich Miescher was the first person to see DNA. He collected used bandages from hospitals. He took the pus from the bandages and used chemicals to break it down. One of the substances he saw was a white, stringy material that came to be called DNA.

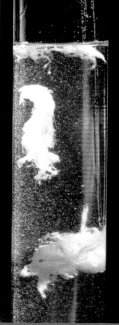

This researcher examines a test tube. The white material in the test tube at right is DNA.

Inventing Plants

Until the 1980s, genetic engineering did not exist. But ten thousand years ago, people were already developing new kinds of plants. They were farming.

Before people began farming, they gathered food from plants wherever the plants happened to grow. These gatherers took whatever they could find to eat. Once people began farming, they could be choosy. They could select and plant the seeds of only the best-tasting, most nutritious plant foods. They could also breed two slightly different plants together to create a third, even better plant. Such improved plants are called hybrids.

These early farmers knew nothing about genes. But they still selected only the plants with the very best genomes to pass along from one generation to another. The process of choosing plants to raise is called selective breeding. It has the same goal as genetic engineering. It just takes longer to get results. Thousands of years of selective breeding have gone into nearly every plant food you eat.

In the 1980s, scientists began using genetic engineering on plants. Genetic engineers can do what selective breeders never could. They can take a gene from one species (specific kind) and add it to the genome of another, and they can do it fast. Plants with one or more genes from another species are called genetically modified, or transgenic, plants.

In the 1800s, American farmers used selective breeding to choose the best corn to plant.

Peanut Problem

Healthful foods can be dangerous to people with food allergies. For example, some people are so allergic to peanut butter that a single bite can send them to the hospital. Scientists are engineering peanuts to remove the allergy-causing genes and make peanut butter safe for everyone.

Corn That Fights Back

Genetic engineers sometimes create transgenic plants to solve problems. For example, corn farmers have serious problems with insects called corn borers. These pests eat and destroy huge numbers of corn plants. Treating fields with chemical pesticides kills these pests, but pesticides cause other problems. They're expensive to buy, they take time to apply to fields, and they can harm the environment. Farmers would be better off with a corn plant that can protect itself from corn borers.

Genetically modified *Bt* corn sprouts in an Iowa field (*top*). *Bt* corn protects itself from the corn borer (*above*).

Genetic engineers found a solution in bacteria known as *Bt*. Bacteria are single-celled, microscopic creatures that live everywhere on Earth—even in your body. *Bt* bacteria live in soil. They produce a poison that kills corn borers. When *Bt* is sprayed on a field, corn borers eat poison along with the corn plants. The poison stops the corn borers from eating more corn. Then the pests starve to death.

To engineer the corn, genetic engineers needed a way to move the pest-killing gene from the *Bt* bacteria to the corn plant genome. Bacteria do not have a nucleus, but they still have a genome. First, the scientists used chemicals called enzymes to cut the pest-killing gene out of *Bt*'s genome. Once scientists had the gene, they made many, many copies of it.

To move the gene into the corn plant's genome, scientists used a gene gun. The gun's tiny "bullets" are made of metal, often gold. Scientists covered these golden bullets with the copies of the pest-killing gene. Then they fired the gene-covered bullets into corn cells.

Once the *Bt* gene entered the corn cell, it became part of the DNA in the corn's genome. The genetically modified cells grew into corn plants. When corn borers ate these new corn plants, they died. *Bt* corn is great news for farmers and bad news for pests.

Shrinking Watermelons

Farmers and scientists continue to use selective breeding to change plants. Petite watermelons come from selective breeding. (If you speak French, you know that petite means "small.") Scientists spent ten years developing petite watermelons.

After every harvest, scientists planted seeds from only the smallest watermelons. Year by year, each new crop of melons was a little bit smaller than that of the year before. A normal watermelon weighs about 20 pounds (9 kilograms). The petite variety weighs about 6 pounds (3 kg). It first appeared in supermarkets in the summer of 2003.

A petite watermelon (left) is one-third the size of an ordinary watermelon (right).

Rescuing Papayas

Papayas are sweet tropical fruits that grow on trees. In the mid-1990s, a deadly virus was destroying Hawaii's papaya crop. The virus slowly spread across the island's papaya plantations (farms), killing the trees.

Before tackling this problem, scientists reviewed what they knew about viruses in humans. An example of a virus is chicken pox. When the chicken pox virus infects your body, you get sick. Your body has to fight off the virus for you to recover. But if you have been vaccinated for chicken pox (usually by getting a shot at the doctor's office), you can avoid getting the disease.

The vaccine gives you weakened copies of the chicken pox virus. These weakened copies warn your body about the disease. Your body develops chemical defenses against it without actually getting sick. The next time the chicken pox virus attacks, your body can fight off the virus with these defenses.

To learn about the virus attacking papaya trees (above) in Hawaii, scientists first studied human viruses like chicken pox (right).

FUN FACT!

The papaya tree can grow from seed to a 20-foot (6-meter) tall tree in less than a year and a half.

Scientists have engineered a papaya tree (*above*) that has built-in resistance to the deadly papaya ringspot virus.

In other words, you are immune to the virus and won't get sick.

To fight the papaya virus, genetic engineers created a vaccine for papayas. They used an enzyme to cut a gene from the deadly virus—a gene that would not make the papayas sick. Then they used a gene gun to shoot copies of this virus gene into papaya cells. The added virus gene worked in papayas the way a chicken pox vaccine works in humans. It immunized the papayas. The researchers grew papaya trees from these immunized cells. The farmers replaced their sick trees with the transgenic papayas. When the virus attacked, the plants did not get sick. Thanks to genetic engineering, Hawaii's papaya crop was saved.

Plant Pharms

Food plants come from farm fields. Medicines are sold in pharmacies. A new word, pharming, is used to describe the engineering of plants or animals to make medicines. Scientists want to engineer plants that can make vaccines for people. Instead of protecting plants—like the papayas—against a disease, these vaccines would protect people against a disease.

Usually, a person gets a vaccine by getting a shot. But some vaccines are oral. An oral vaccine is a pill, capsule, or liquid (or fruit or vegetable!)

that you put in your mouth. Ordinary vaccines are expensive to make, and they have to be refrigerated. In poor areas of the world without electricity, doctors don't have refrigerators so they can't keep the vaccines cold. But vaccines in fruits or vegetables would allow people anywhere in the world to be vaccinated.

One group of scientists is engineering bananas to deliver an oral vaccine for hepatitis B, a virus that harms the liver. Then people could be protected from this disease just by peeling and eating a banana. These bananas would not be for sale in the supermarket. Only doctors and nurses would be able to give them out.

Children around the world may one day be able to eat bananas that protect them from disease.

Making the Earth Safer

Land mines are small but powerful explosives that people hide just below the ground during wars. When a person steps on a land mine, it explodes. The person may be seriously injured or killed. A land mine may stay hidden for years, long after a war is over, until someone accidentally sets it off.

Scientists are engineering plants to make them do something no plant has ever done before. The plants will change color or glow when explosives are nearby. Scientists want to grow these plants in areas with large numbers of land mines. Then explosives experts could easily find the mines and safely remove them before the mines hurt anyone.

This soldier defuses a land mine in Bosnia.

Weed and Allergies

Even though transgenic plants can do a lot of good things, some people are worried. They fear that new plants could have unintended consequences. These are results—especially negative results—that we don't expect.

Farmers use transgenic crops to fight pests and weeds. Some crops have been engineered to survive when they are sprayed with herbicides. These chemicals kill weeds. But sometimes genes can spread from one plant to another. What if genes that protect crops from herbicides spread to the weeds? The result could be "superweeds" that no herbicide could kill.

Transgenic foods and drinks might have unexpected effects on people too. Such products as bread, potato chips, pizza, and soda pop all have transgenic ingredients, such as wheat, potatoes, tomatoes, or corn. What if some people are allergic to transgenic foods? Millions of people might become sick. Scientists and governments do many tests to make sure foods are safe, but people still worry that in the long run, the transgenic foods will prove to be unsafe. What do you think?

chapter 3

Improving Animals

For thousands of years, people have used selective breeding to make better animals. In 1791 Massachusetts farmer Seth Wright found some short-legged sheep in his flock. He kept breeding the sheep with the shortest legs together until he had a whole flock of short-legged sheep. These sheep couldn't jump over his fences and escape!

This short-legged ram (right) and long-legged ewe (left) produce offspring with very short legs (center).

Genetic engineering of animals is more difficult than genetic engineering of plants. Animals reproduce by laying eggs or giving birth to live babies. To make a transgenic animal, scientists must change the genome of an embryo. An embryo is an animal in the earliest stage of life—it's just a few cells that will eventually grow into a whole animal. After engineering the embryo, scientists put it into a female animal that will give birth to their creation.

Inventing Super-Mouse

One of the first transgenic animals scientists created was a supersized mouse. To make it extra large, scientists gave it growth genes from a rat. Here's what they did:

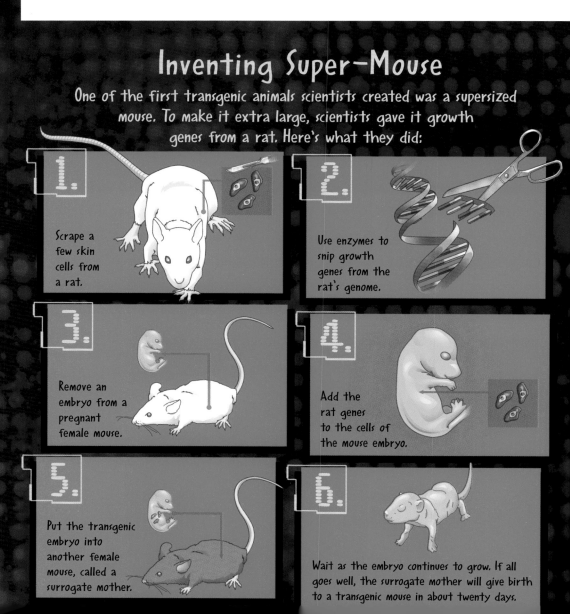

1. Scrape a few skin cells from a rat.

2. Use enzymes to snip growth genes from the rat's genome.

3. Remove an embryo from a pregnant female mouse.

4. Add the rat genes to the cells of the mouse embryo.

5. Put the transgenic embryo into another female mouse, called a surrogate mother.

6. Wait as the embryo continues to grow. If all goes well, the surrogate mother will give birth to a transgenic mouse in about twenty days.

Fast Fish

Many of the fish we eat do not come from oceans and rivers. They are raised in tanks or ponds on fish farms. Most farmers want their products to grow quickly. The faster something grows, the faster they can sell it and the more money they can make.

For example, ordinary catfish stop growing in winter, when the genes that tell the fish's body to grow turn off. If catfish kept growing all year-round, fish farmers could breed more of them.

Scientists know that some kinds of salmon, carp, and zebra fish grow all year long. So they inserted growth genes from these fish into catfish genomes. The experiment worked. The modified catfish grow all year, and catfish farmers can produce fish faster than ever.

Ordinary catfish (*top*) do not grow all year-round. Scientists are using different techniques, including genetic engineering, to produce fast-growing fish (*above*).

Transgenic fish are not sold in supermarkets yet, though. The U.S. government is still testing them to be sure they are safe for everyone to eat.

Pig Pollution

Farmers who raise animals must find ways to get rid of the animals' manure, or waste. They often collect the waste and sell it as fertilizer. Fertilizer helps plants grow bigger and faster. Pig manure makes good fertilizer, but it also contains a chemical called phosphorus that harms the environment. Pig feed has a lot of phosphorus in it. Since pigs cannot digest all of the phosphorus, it ends up in their manure.

Genetic scientists decided to create pig feed with an enzyme that helps pigs digest more phosphorus. But the enzyme broke down in the feed, before the pigs could ever eat it.

Then genetic engineers decided to create pigs that could make their own enzyme to break down phosphorus. They used genes from bacteria and mice to create a brand-new gene for pigs. They added the new gene to pig embryos, put the engineered embryos into female pigs, and waited for the piglets to be born.

Fatal Phosphorus

Why is the phosphorus in pig manure bad for the environment? Pig manure is used as fertilizer on farm fields. Plants can't absorb all of the phosphorus in this fertilizer. Rainwater washes the extra phosphorus into rivers and lakes, where it speeds up the growth of algae. These small, floating plants use up lots of oxygen. When the oxygen level in water falls too low, fish can't breathe, so they start to die off.

Phosphorus helps algae to grow in lakes and ponds.

These new transgenic pigs can digest as much as 75 percent more phosphorus than ordinary pigs. The transgenic pigs' manure has less phosphorus in it. Because their manure is less harmful to the environment, these transgenic animals are known as Enviropigs.

Canadian scientists have engineered pigs with cleaner manure.

Silk Milk

Ounce for ounce, spider silk is one of the strongest materials on Earth. It's five times stronger than steel. When spider silk is woven into fabric, it can stop a speeding bullet. But spider silk is very hard to get in large amounts. Spiders can't be raised on spider silk farms. When spiders are kept close together, they fight and kill each other.

So genetic engineers went to work. They snipped out the spider genes that call for silk and engineered them into the genomes of female goat embryos. These new transgenic goats produce milk that contains lots of the proteins that are used to make spider silk. The silk proteins are collected from the milk, processed, and spun into fibers. These fibers can be used to make extremely strong but lightweight cloth. One day soldiers may actually wear bulletproof vests made from spider silk!

FUN FACT!

The toughest spider silk is the dragline silk from the orb weaver spider. Dragline silk must support a spider's weight, so it is the strongest silk a spider makes.

Heart of a Pig

Some animal genetic engineering projects sound more like science fiction than science fact. In the United States, patients have to wait for months or even years to get an organ transplant. Scientists would like to supply healthy new organs, such as hearts, lungs, kidneys, and livers, to all people who need them.

One experiment is engineering pigs to produce human body parts. Scientists have engineered the genomes of these pigs so that their organs are more like human organs. They hope that human bodies will accept these new body parts. But scientists still have a lot more experimenting to do before they will be ready to replace human organs with pig organs.

Humans who need new hearts must have another human heart (above) transplanted into their bodies. One day, pig hearts instead of human hearts may be transplanted into humans.

Fun Fish

Not all transgenic animals solve a problem. Some are used just for fun. One example is a transgenic zebra fish that glows in the dark. It was originally invented to help detect pollution by glowing only in polluted water. But by accident, a scientist created several varieties of fish that glow all the time. The glow comes from the genes of other sea creatures. Zebra fish with a sea anemone gene glow red. Those with a jellyfish gene glow green.

Fluorescent genes make these small fish shimmer and glow in the dark like neon signs. They are sold under brand names such as "GloFish," "Night Light Fish," and "Night Pearls." They are the world's first transgenic house pets.

Genetic engineers have created zebra fish that glow.

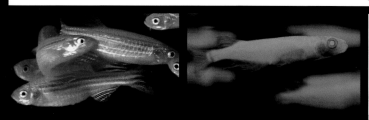

Cloning Animals

Scientists have invented another new kind of animal, a clone. Animal cloning is not truly genetic engineering. Scientists don't change the animal's genome. But cloning uses many of the same tools and methods as genetic engineering.

What makes a clone different from other animals? Most of the time, an animal has two parents. Its genome is a mixture of genes from its mother and its father. But a clone's genome is an exact copy of the genome of only one parent.

Why clone animals? Imagine if you could clone copies of your very best dairy cow, the one that gives the most and best milk. Or the pig that produces the best pork.

Dolly, the first cloned sheep, lived from 1996 to 2003.

Or the sheep in your flock that produces the very best wool. Then every animal you owned would be the very best of its kind.

The very first cloned mammal was Dolly the sheep. She was born in 1996. A team of scientists from Scotland cloned her from an adult female sheep.

Hello, Dolly

See if you can answer this question:
Which of the sheep
(#1, #2, or #3)
is Dolly's parent?

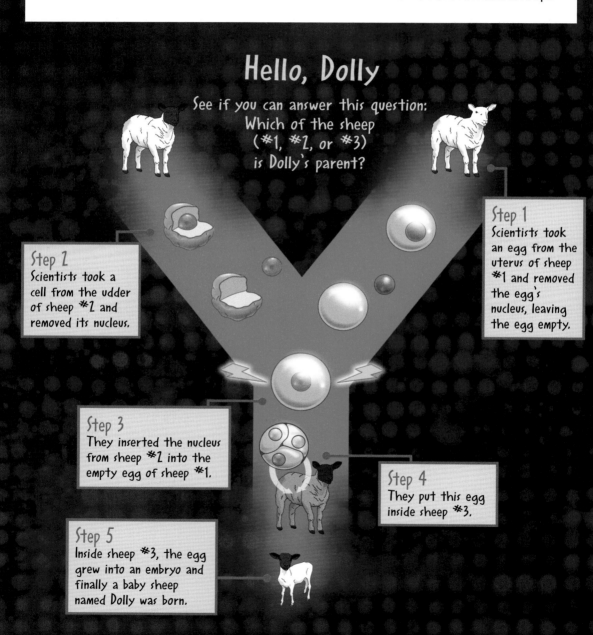

Step 1
Scientists took an egg from the uterus of sheep #1 and removed the egg's nucleus, leaving the egg empty.

Step 2
Scientists took a cell from the udder of sheep #2 and removed its nucleus.

Step 3
They inserted the nucleus from sheep #2 into the empty egg of sheep #1.

Step 4
They put this egg inside sheep #3.

Step 5
Inside sheep #3, the egg grew into an embryo and finally a baby sheep named Dolly was born.

Dolly is a clone of sheep #2 because sheep #2 supplied the nucleus that contained Dolly's genome.

Unintended Consequences

Some people fear that creating transgenic animals will have unintended consequences. They worry that the animals may not turn out exactly as planned. But other people argue that genetic engineering has the same result as selective breeding—only faster.

Genetic engineers want to eliminate malaria, a virus carried by mosquitoes that kills millions of people each year. Scientists are trying to engineer mosquitoes that resist the virus. Insects that don't carry the virus can't pass it on to humans. Scientists hope to release thousands of engineered mosquitoes into the wild, where they can breed with other mosquitoes. If all goes well, the antimalaria genes will spread through the whole mosquito population, and malaria will be conquered.

Hold on a minute, say people who worry about genetic engineering. What if these transgenic mosquitoes carry other diseases that we don't even know about yet? Then we may create worse problems than the one we solve.

Genetically engineered animals, including mosquitoes and salmon, raise concerns.

Some people have the same kinds of worries about "supersalmon." These transgenic salmon are engineered to grow larger than ordinary salmon. They are raised on fish farms and have never been in the wild. But suppose they escape into rivers and oceans and breed with wild salmon? Transgenic salmon have never had to fight off predators and hunt their own food. This new breed of transgenic-wild salmon may not be able to survive in the wild. Then the entire supply of wild salmon could be killed off. What do you think?

Engineering People

he human genome has more than twenty thousand genes. Your genome holds all the instructions for making your body and keeping it running. No one is born with a perfect genome. Everyone has some genes that don't work as well as they should. These are called faulty genes. As people get older, more faulty genes show up.

Your body has trillions of cells. A trillion is a million million. Every day, old cells die and new ones replace them. Each

These human skin cells, shown through a microscope, have been stained with dye to make their structures visible.

new cell is supposed to have the exact same copy of the genome. But sometimes things go wrong. A gene doesn't get copied exactly right. These mistakes are called mutations.

Most mutations are harmless. Some even help make living things healthier and stronger. But sometimes harmful mutations show up. Some mutations cause deadly diseases. For example, one mutation causes the disease severe combined immunodeficiency (SCID). This condition makes people unable to fight off ordinary colds and other infections. These infections can kill them.

To cure diseases like SCID, scientists have recently started using genetic engineering to repair faulty genes. Doctors and genetic researchers are learning how to get into human genomes and replace damaged genes with healthy ones. This is known as gene therapy.

Ashanthi DeSilva (right) and Cynthia Cutshall (left) were the first two people ever to receive gene therapy.

Viruses That Help

In gene therapy, scientists do a lot of the same things they do to create transgenic plants and animals. But the human genome is harder to engineer. Most of the time, scientists want to change the genome of a person who has already been born. The trickiest part of gene therapy is delivering engineered genes to the genome.

Scientists sometimes use viruses for delivering genes. Scientists know how to disable, or weaken, viruses. They get rid of the parts of a virus that cause sickness. Scientists then put healthy genes into the disabled viruses. Then the viruses can deliver the healthy genes into the genome. If all goes well, the healthy genes replace the faulty genes and the patient is cured. This type of therapy has been used to help SCID patients.

Stem Cells to the Rescue

Your body has different types of tissues, such as muscle, lung, skin, and heart. Each tissue type has its own special repair crew of cells called adult stem cells. (Even in newborn babies, these cells are called adult stem cells.) For example, when you cut your finger, your body sends out a crew of skin stem cells. This repair crew travels to the cut and makes new skin to heal it.

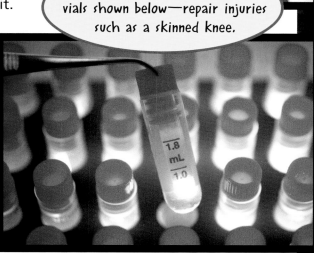

Stem cells—contained in the vials shown below—repair injuries such as a skinned knee.

These repair crews can't always handle the job when a person is very sick. Someone with heart disease, for instance, might need a huge supply of heart cells to repair the damage—more than the stem cell repair crews can possibly supply.

In a case like this, scientists are hoping that it will be possible to take a few heart stem cells from the patient and clone, or copy, them. They would inject many copies back into the patient. With more stem cells, the patient's body would be able to do a better job of healing itself.

FUN FACT!

Your body makes new skin cells all the time. You shed your skin, little by little, every thirty-five days. That means you grow a completely new skin about ten times a year!

Scientists also believe they could make stem cells grow into whole body parts. Instead of using pigs to make human organs, scientists could grow human organs from human stem cells. In a laboratory, scientists would take stem cells from a patient and use them to make all of the cells of a new organ. Doing this would help people who need organ transplants. It could also help healthy people live longer. As people grow older, their organs naturally wear out. Before they do, the old organs could be replaced with brand-new ones grown from stem cells.

Naked Mice for Sale!

Laboratories supply different kinds of mice to scientists for research. One variety is known as a nude mouse (below). The genes that tell the mouse's body how to grow hair are missing from its genome.

Going against Nature?

People who worry about genetic engineering worry the most about engineering humans. Some people believe that genetic engineering is too risky to try on humans.

For example, patients with SCID have been helped by gene therapy. But two French boys with SCID who were being treated with gene therapy came down with leukemia, or cancer of the blood. The boys' cancer was successfully treated, but it worried scientists. Scientists believe that the gene used to make the boys healthy ended up in the wrong part of their genome and that this caused the leukemia. When the two boys became sick, other genetic engineering experiments around the world were stopped.

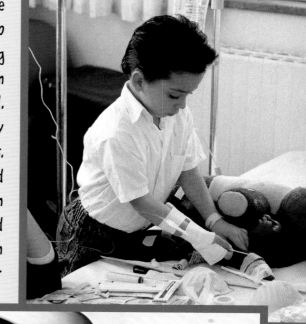

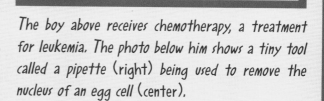

Some people object to gene therapy and growing human organs in animals because scientists don't know enough about how the human genome works. These people say we should wait until we know more. Other people object because they believe changing genomes

The boy above receives chemotherapy, a treatment for leukemia. The photo below him shows a tiny tool called a pipette (right) being used to remove the nucleus of an egg cell (center).

goes against nature. Human and animal genomes were not meant to be invaded by scientists and changed, they argue. What do you think?

Knockout Mice

Scientists have a lot to learn about the part that each gene plays in the human body. Laboratory experiments on people can be very dangerous. If an experiment on a person did not work as expected, the person could suffer and die. So to learn more, scientists experiment on animals.

Mice are popular lab animals. They are easy to handle, they reproduce quickly, and they have many of the same genes that humans have. To test different genes, scientists go into the genome of a mouse embryo and knock out, or turn off, a specific gene. The mouse is called a knockout mouse. Once it is born, researchers watch closely to see how the mouse is different from mice that are not missing the gene. These experiments teach scientists what part the knocked-out gene plays in growing and running the body.

Mice have a lot in common with humans, even though they have forty chromosomes and humans have forty-six.

FUN FACT!

One female mouse can have more than one hundred babies in one year.

Genetic engineering has the potential to change our lives in many different ways. Let's look at some of the wildest ideas that genetic engineers hope to bring to life in the future.

Blue Roses

Roses are red. They are also pink, yellow, white, and purple. But roses are not blue—not yet. Many people want to be able to buy blue roses and grow them in their gardens. No one has ever been able to breed a blue rose. But genetic engineering may change this.

Roses come in hundreds of varieties and colors, but none of them are blue.

Companies that sell flowers have spent millions of dollars to engineer a blue rose. Scientists have tried engineering genes from blue petunias and other flowers into rose genomes. So far the results are disappointing—and purple. But the company that can create blue roses will make lots of money. So scientists will keep trying to engineer them.

Engineered Pets

Are you allergic to cats? Many people are. People with this allergy suffer itchy eyes, scratchy throats, runny noses, and violent sneezing when they are around cats. But some of these people love cats and wish they could have one for a pet.

Scientists have found the cat genes that make people allergic to cats. Tests show that cats could live normal, healthy lives if these genes were engineered out of their genome. Scientists are trying to engineer allergy-free cats to sell in the United States and Japan.

Nonallergenic cats would likely be created by cloning. The cats above are two of the world's first cloned cats.

Members of animal rights groups (organizations that work to prevent cruelty to animals) say the tests scientists perform will harm cats and kittens. They fear that too many animals will be harmed to make the results worthwhile.

Healthy Couch Potatoes

Scientists have engineered mice to be terrific long-distance runners. The key is "fat-switch" genes. When fat-switch genes are turned on—in mice or in people—the body burns fat. Scientists have engineered mice so their fat-switch genes are always turned on. This gives them a constant supply of energy. The engineers tested the mice by having them run on treadmills. The "marathon mice" could run nearly twice as far as other mice before stopping.

Someday, scientists say, they may be able to engineer people the same way. With their fat-switch genes turned on all the time, some people might turn into super marathon runners, just like the mice.

But this is not why scientists have engineered fat-switch genes. They want to help people lose weight. The engineered mice stay thin even when they eat only high-fat foods and don't exercise. Their fat-switch genes keep the mice's bodies burning fat even when they're fast asleep!

Scientists use exercise wheels to test how long mice are able to run.

Super Baby

Picture a four-year-old boy holding a 7-pound (3 kg) weight in each hand while holding his arms straight out at his sides. Many full-grown adults aren't strong enough to do this. The boy, who lives in Germany, has a rare genetic mutation that keeps his myostatin gene permanently switched off. Myostatin is a protein that limits muscle growth. So the boy has twice the muscle of other kids his age and only half the body fat. A rare breed of cattle, Belgian Blue, has the same kind of mutation. These powerful cattle have bulging muscles all over their bodies and hardly any fat.

Researchers are working on gene therapy treatments for humans based on this mutation. They want to engineer people so that their myostatin gene is switched off. But they don't want to create more superstrong people.

Adult male Belgian Blue cattle, like the one shown below, weigh approximately 2,700 pounds (1,200 kg).

They want to treat people who have muscular dystrophy. This disease weakens muscles and can cripple the bodies of children and adults. With the myostatin gene switched off, people with muscular dystrophy might be able to repair their own damaged muscles.

Cures from Chips

You've heard of computer memory chips. These tiny slivers store huge amounts of information. Recently scientists have created tiny glass gene chips for storing information about diseases. A doctor can take a sample of a patient's cells and coat the gene chip with them. A single chip shows all the patient's genes. When a gene is switched on, it lights up on the chip.

These images show gene chips. The areas of the chip that light up indicate genes that are switched on.

The chip gives a picture of a patient's genome. If the scientists know that a patient has a certain disease, they can compare the gene chip with the gene chips of many patients who have the same disease. Then scientists can identify which genes may be causing the disease. They can work on designing medicines and gene therapy treatments for patients.

Designer Babies

Imagine that we know how to engineer embryos to produce "perfect" children. Parents could walk into a doctor's office and design the genome of their child-to-be. Let's say their idea of a perfect child is a girl with black hair and brown eyes who will be 5 feet 10 inches (1.8 meters) tall when she's an adult. They want her to have special talents for doing math and playing the piano. And they think her personality should be just a little shy.

IVF

Scientists may still be years away from creating designer babies or human clones, but they have already figured out one important piece of that puzzle. In vitro fertilization (IVF) was developed in the 1970s to help infertile couples have children. Scientists use a couple's eggs and sperm to grow embryos in lab dishes. Some of these embryos are placed inside a woman to grow into a baby. IVF has resulted in the birth of more than one million babies worldwide.

As part of in vitro fertilization, a pipette containing sperm (left) is inserted into a human egg cell (right).

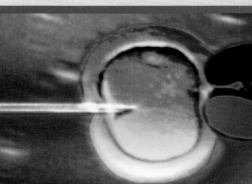

Whoa, Not So Fast!

Many people oppose genetic engineering. They say scientists are moving much too quickly. They believe that scientists don't know enough about genomes yet to be sure their work won't end up doing more harm than good.

This goes for engineering plants and animals as well as humans. Opponents admit that, so far, no transgenic plant or animal has harmed anyone. But no matter how careful we think we are, they say, it may not be careful enough.

Other people, including many scientists, disagree—sort of. No one says that nothing bad could ever happen. Scientific studies warn that transgenic plants and animals will likely bring surprises in the future. And some of these surprises may bring problems. But that's how it is with anything new, these people say. We can never be totally sure that a new technology is completely safe. We must use it, see what happens, and keep an eye out for unintended consequences. We must move ahead with the genetic engineering of plants and animals, they say—but cautiously. What do you think?

A researcher examines genetically modified corn (top). Protesters in Germany (above) want their government to prohibit genetically modified food.

FUN FACT!
SOMETHING TO THINK ABOUT—
How would you feel if one day you learned that your parents had chosen your appearance, personality, and talents for you, before you were born?

Could scientists ever really engineer a genome in the embryo to grow into this ideal child? Most experts agree that someday we will design genomes for physical features, such as hair and eye color and height. But they doubt that we could ever engineer talents and personality. There is no such thing as a math gene or a shyness gene, as far as scientists can tell. And talents and personality grow little by little, day by day. They don't just show up all at once. Still, some experts insist that designer babies of some kind will exist one day.

Human Cloning

All sorts of animals have been cloned by scientists: sheep, mice, cows, pigs, cats. But not humans. Not yet.

To clone a human, scientists would run through the same steps it took to clone Dolly the sheep (see page 26). They would put the nucleus from the parent's cell into an empty egg, put the egg into a woman, and wait for the egg to grow into a live human baby. Clones share the same genome, so parent and child would look a lot alike. But each one would still have his or her own special personality.

Scientists have tried creating human clones in a laboratory. So far no one has succeeded. But experts say that eventually a human likely will be cloned and more will follow.

Many people strongly oppose human cloning and designer babies. They believe creating life should be left to Mother Nature. Each child's

These identical triplets are human clones that occurred naturally.

genome should be new and unique, a never-before-seen combination of the parents' genomes, as nature intends.

Some scientists disagree. They say that parents already mold and shape their children in many different ways. They give them piano lessons and send them to computer and basketball camps. They give them advice on what to wear and eat and whom their friends should be. So why shouldn't parents be allowed to shape their children's genomes too? What do you think?

FUN FACT!
Have you ever met identical twins? Then you know human clones. Identical twins come from the same fertilized egg. So do identical triplets and quadruplets!

Glossary

adult stem cells: cells that can become any kind of tissue the body needs

bacteria: single-celled living organisms

cell: the basic unit of any living organism. Cells carry on the processes of life, according to instructions they get from genes.

chromosomes: rod-shaped structures in the nucleus of most cells that contain genes. Humans have twenty-three pairs of chromosomes.

clone: an exact copy of a gene, of a whole cell, or of a complete organism

designer babies: babies created through genetic engineering whose genes have been changed according to the parents' wishes

DNA (deoxyribonucleic acid): the molecule in the nucleus of most cells that holds genetic information passed from parent to child during reproduction

embryo: a developing creature in the very early stages of life

enzymes: proteins that makes reactions occur more quickly or efficiently

gene: a unit of chemicals in the genome that holds instructions for how to make one or more proteins

gene therapy: an experimental procedure to repair damaged genes or replace them with healthy genes

genetic engineering: changing genes to create new organisms

genome: all the genetic material in the chromosomes of an organism

knockout mouse: a mouse used in genetic engineering experiments. Scientists knock out, or disable, one or more genes to find out what instructions those genes carry.

mutation: a random, accidental change in an organism's DNA. Mutations can be harmless, harmful, or helpful.

nucleus: the central part of a cell that contains the genome

pharming: engineering plants or animals to make medicines

protein: a molecule created according to instructions from a gene. Proteins carry out all the processes of life, from making the hair on your head to breaking down food in your stomach.

selective breeding: breeding only the very best plants or animals to produce superior offspring

transgenic: having genes from more than one species. An organism is transgenic when one or more genes from another species have been engineered into its genome.

vaccine: a preparation of weakened viruses or bacteria that helps build up the body's defenses against diseases and that immunizes, or protects, an organism from disease

virus: a tiny organism that produces a disease by copying itself in another organism's cells

Selected Bibliography

Aldridge, Susan. *The Thread of Life: The Story of Genes and Genetic Engineering.* Cambridge, UK: Cambridge University Press, 1996.

Bryson, Bill. *A Short History of Nearly Everything.* New York: Broadway Books, 2003.

Hubbell, Sue. *Shrinking the Cat: Genetic Engineering before We Knew about Genes.* Boston: Houghton Mifflin, 2001.

Stock, Gregory. *Redesigning Humans: Our Inevitable Genetic Future.* Boston: Houghton Mifflin, 2002.

Further Reading and Websites

Balkwill, Fran. *DNA Is Here to Stay.* Minneapolis: Carolrhoda Books, Inc., 1993.

Cefrey, Holly. *Cloning and Genetic Engineering.* New York: Children's Press, 2002.

Fridell, Ron. *Decoding Life: Unraveling the Mysteries of the Genome.* Minneapolis: Lerner Publications Company, 2005.

Marshall, Elizabeth L. *High Tech Harvest.* Danbury, CT: Franklin Watts, 1999.

Nardo, Don. *Cloning.* San Diego: Lucent Books, 2002.

Seiple, Samantha, and Todd Seiple. *Mutants, Clones, and Killer Corn: Unlocking the Secrets of Biotechnology.* Minneapolis: Lerner Publications Company, 2005.

Snedden, Robert. *DNA & Genetic Engineering.* Chicago: Heinemann Library, 2003.

Torr, James D., ed. *Genetic Engineering: Opposing Viewpoints.* San Diego: Greenhaven Press, 2001.

DNA from the Beginning

http://www.dnaftb.org/dnaftb/

This comprehensive website uses text, images, and animation to explain the basics of genetics and genetic engineering.

Ethical, Legal, and Social Issues—Genome Research

http://www.ornl.gov/sci/techresources/Human_Genome/elsi/elsi.shtml

This section of the Human Genome Project's website discusses ethical issues. It includes articles about privacy, gene therapy, genetically modified foods, and also includes links to other websites.

Guardian Unlimited Picture Gallery

http://www.guardian.co.uk/gall/0,8542,627251,00.html

This website showcases photos and basic information about recently cloned animals.

Photo Acknowledgments

The images in this book are used with permission of: Mitch Doktycz, Life Sciences Division, Oak Ridge National Laboratory; U.S. Department of Energy Human Genome Program, <http://www.ornl.gov.hgmis>, background image throughout and p. 38; © Jim Richardson/CORBIS, p. 4 (left); © Rob C. Nunnington; Gallo Images/CORBIS, p. 4 (right); © Jeff Albertson/CORBIS, p. 5; © A. Barrington Brown/Photo Researchers, Inc., p. 6; © L. Clarke/CORBIS, p. 8; © AP | Wide World Photos, pp. 9 (top), 16, 21 (bottom); © Clouds Hill Imaging Ltd./CORBIS, p. 9 (bottom); © Maximilian Stock Ltd./Science Photo Library, p. 10 (bottom); © PhotoDisc Royalty-Free by Getty Images, pp. 10 (top), 40 (top); © Hulton Archive/Getty Images, p. 12 (top); © Sam Lund/Independent Picture Service, pp. 12 (bottom), 14; © Kent Foster/Visuals Unlimited, p. 13 (top); © Agricultural Research Service, USDA, p. 13 (bottom); © Douglas Peebles/CORBIS, p. 15 (left); © Lester V. Bergman/CORBIS, p. 15 (right); © Joel W. Rogers/CORBIS, p. 17; © Bojan Brecelj/CORBIS, p. 18; © Herbert Gehr/Time Life Pictures/Getty Images, p. 19; © Nikolas Konstantinou/Getty Images, p. 21 (top); © Sally A. Morgan; Ecoscene/CORBIS, p. 22; © University of Guelph, Ontario, Canada, p. 23; © Bruce Dale/National Geographic/Getty Images, p. 24; © www.glofish.com, p. 25 (upper left and right); © SAM YEH/AFP/Getty Images, p. 25 (upper center); © MC LEOD MURDO/CORBIS, p. 25 (lower right, all); © John Downer/Getty Images, p. 27 (top); © Brandon D. Cole/CORBIS, p. 27 (bottom); © Jim Zuckerman/CORBIS, p. 28; © Ted Thai/Time Life Pictures/Getty Images, p. 29; © SCF/Visuals Unlimited, p. 30 (top); © Peter Macdiarmid/Reuters/CORBIS, p. 30 (bottom); © David A. Northcott/CORBIS, p. 31; © David A. Wells/CORBIS, p. 32 (top); © Getty Images, p. 32 (bottom); © Royalty-Free/CORBIS, pp. 33, 34, 36; © Genetic Savings & Clone, p. 35; © Yann Arthus-Bertrand/CORBIS, p. 37; © Lester Lefkowitz/CORBIS, p. 39; © Johannes Eisele/AFP/Getty Images, p. 40 (bottom); © Paul Barton/CORBIS, p. 42.

Front cover: Mitch Doktycz, Life Sciences Division, Oak Ridge National Laboratory; U.S. Department of Energy Human Genome Program, <http://www.ornl.gov.hgmis>, (background image and right); © Comstock Images (left); www.glofish.com (center); © David A. Northcott/CORBIS, (mouse).

About the Author

Ron Fridell has written for radio, television, and newspapers. He has also written books about the Human Genome Project and the use of DNA to solve crimes. In addition to writing books, Mr. Fridell regularly visits libraries and schools to conduct workshops on nonfiction writing. He lives in Evanston, Illinois.